WARNING
TAKING SHELLFISH FROM
THESE WATERS OR
SHORES IS PROHIBITED
PENALTY FINE OR IMPRISONMENT OR BOTH
RHODE ISLAND STATE DEPARTMENT OF HEALTH

Comparative Marine Policy